AF342000

Insects
The Seasons in Their Lives

Insects

The Seasons in Their Lives

by Beverly Dobrin Wallace

WITH ILLUSTRATIONS BY THE AUTHOR

The Bobbs-Merrill Company, Inc.
INDIANAPOLIS/NEW YORK

ISBN 0-672-51784-1
Library of Congress catalog card number 74-17689
Designed by Victoria Gomez
Manufactured in the United States of America

First printing

To
Henry

Without insects
would we have this flower?
What vegetables would people cook?
How would the world look
if insects didn't help
to make the seed that makes the tree?
There is a peculiar harmony
between the flower and the bee,
between predator and prey.

1

It is winter.
Trees are tall and black, jutting out
of the snow. They are resting now and
waiting for a change in the weather.

A large hornet's nest hangs from a branch
reaching high out over the pond. It is held
by a strong black vine. Heavy, it sways and
swings from the branch. But it is
silent now. No buzzing at the entrance hole.
Below, on the ground, the dried winter
bouquets make ragged designs, black on white.

The goldenrod looks like this.

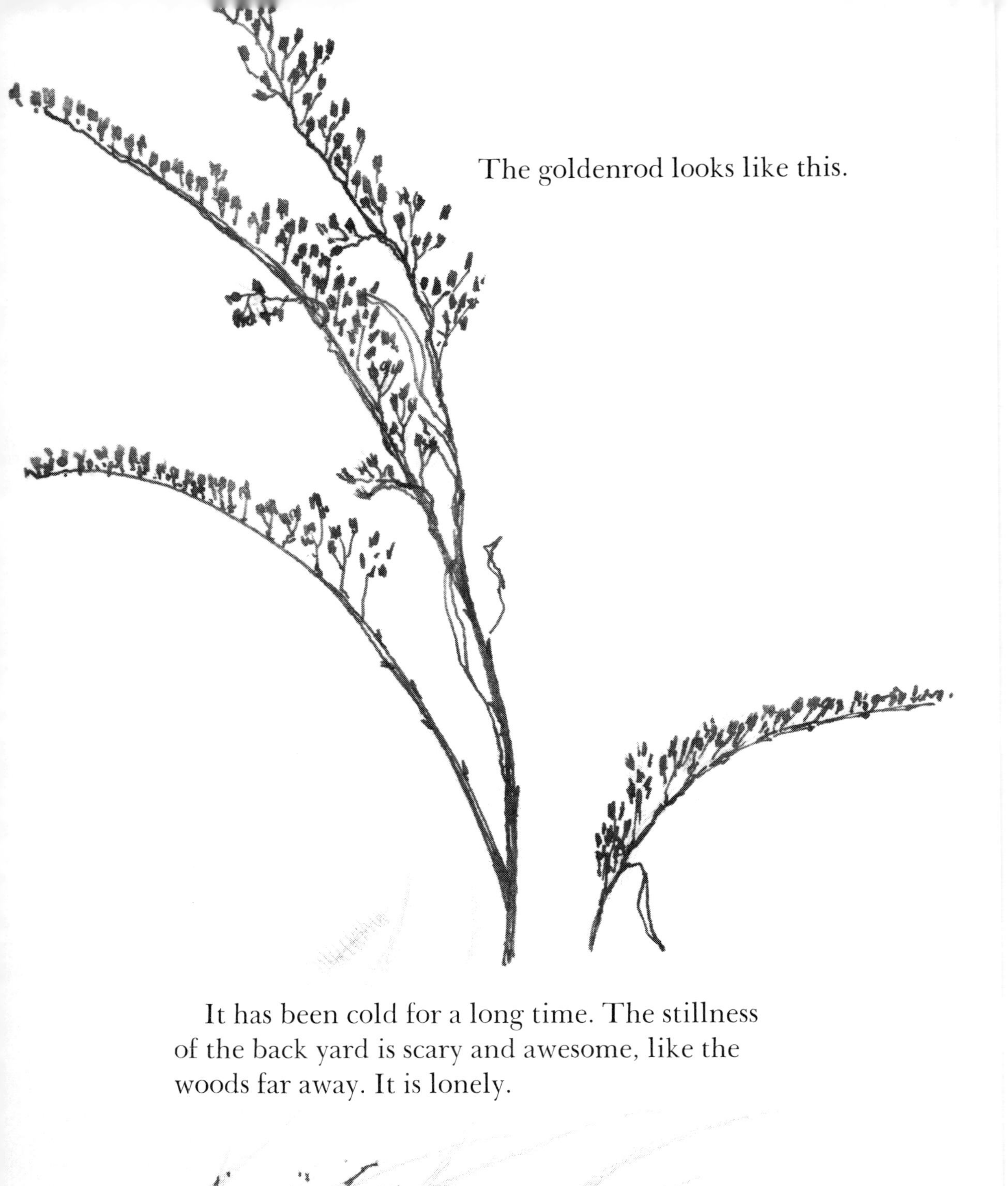

It has been cold for a long time. The stillness
of the back yard is scary and awesome, like the
woods far away. It is lonely.

The Queen Anne's lace
looks like this.

And the grasses are feathery,
dropping their tiny seeds
into the snow.

A honeybee hive is sheltering a queen
and her attendants. No new bees will be born
until spring. The small cluster that rests
inside has nothing to eat except the honey
that was stored in the comb last fall.
While winter winds bluster outside, the
queen sleeps. She sleeps to save her strength
for creating a new brood when spring comes.

Inside a winter beehive

At dawn, against the crisp horizon
a few birds sing, trying their music for spring.

Under the ground on these cold days
two small still shapes lie back to back:
a seed and an egg, each alive yet unborn.
They look alike as well. Each has a shell;
each is white. Both will open in the spring.

11

A larkspur seed and a grasshopper egg quietly
lie in the ground, all winter asleep.
But now the spring gently wakes them, for it's time
to get up! The beetles have plowed the earth;
the rains have softened their dwellings.
The larkspur seed bulges against the warm,
loose earth. Shiny and tight—it bursts!
A little root escapes to find water rich with food
for the baby plant. A shoot finds its way up to the
light. Tiny swellings will become leaves.

The grasshopper egg has spent the winter
underground alongside many other grasshopper
eggs in a warm, waterproof eggpod. The creatures
inside have been changing into little grasshoppers.
One of them feels the new season in the earth.
It cracks the shell with its head and wriggles up
through the grains of soil, out into the sunlight.
Limp, it rests on the ground, gaining strength,
its body hardening. Then it hops to the
top of a stalk of grass. Gazing down,
it regards the busy crowd below,
all hatching from the eggpod.
At night it huddles with the other
young grasshoppers under clover and
larkspur leaves. When the warm
day comes they all gobble plants.

Grasshoppers love the sun, love
green things springing up all around.
They love the taste of meadow grass.
The larkspur plant is for nibbling, and
strawberry leaves are delicious. The
grasshopper eats and eats, and eats some
more, and grows and grows—but its skin
does not. So it splits the skin and
leaves it on a twig. The skin looks like
its ghost, just sitting there. The
grasshopper lives in its new skin and
grows and grows. Five times this
will happen, leaving five little ghosts,
each bigger than the last, each
made of a skin that is not just a
skin but an outside skeleton.

The larkspur plant grows taller. Its
stem towers over broad leaves below.
Along the stem small shoots break
through. They will hold buds,
treasures wrapped in a green sepal.
The sepal is the ring of petals
that protects the bud before it
opens. Each sepal forms a cup to
carry a blossom and give it
its own little vase.

Now the larkspur flower must wait
for its partner, the bumblebee.

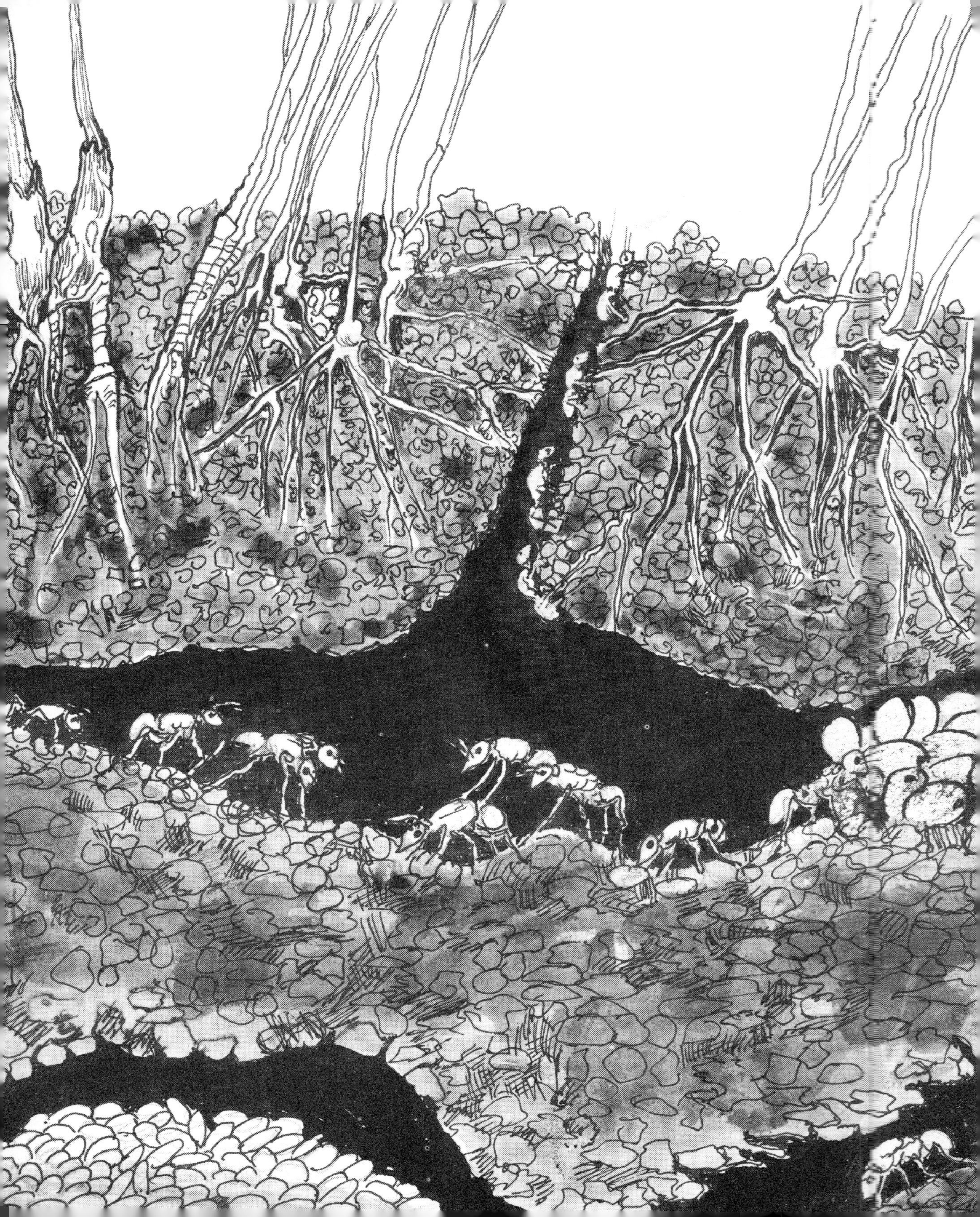

Spring!

The damp earth crumbles. The pond comes
alive with lily pads and dragonflies. Deep
underground, in tunnels below the frost line,
is a secret dwelling. A small colony of
ants has survived the winter in this dark
place. Here they tended aphid eggs
collected in the fall and brought below
for safekeeping. Aphids are very tiny insects.

The dark chambers are getting warmer.
Seeds are sprouting roots over all the
garden. It is time to carry the aphid eggs
up the long passageway to the roots, where
they will hatch. Baby aphids will suck
sap from the plants and make honeydew, which
the ants love to eat.

Tiny ant eggs, licked and tended by nurse
ants, are about to hatch into ant babies.
Some have already hatched. They look like
worms with a mouth. We call them larvae.
When they grow big enough they will spin
cocoons in which to sleep. We call these
pupae. While they sleep, their bodies
change into adult ants.

17

Cocoons are piled together. When a cocoon
smells a special way, the nurses know that the
ant inside has awakened. They split the cocoon
to let it out. Now the ant colony is larger.
But there is plenty of work for all.

A mother robin lands and the earth rumbles.
One of the ants pokes out of a hole in the
soil to look around. It climbs a tall thick
blade of grass for a better view. It sees
the robin listening sideways to the ground.
Swiftly she pierces the earth with her beak,
and swiftly she flies off with a worm for
her babies. The little ant climbs down the
grass and joins its companion to search for meat.
Meat is good for ant babies and is a treat for
all the ants in the colony.

A spring cankerworm hangs by a thread from the leaf it was eating. Carefree, it turns a somersault. One, two, three; at each end an ant grabs it. The cankerworm was not expecting this. Wriggling and turning, it tries to escape from the ants' jaws. They drag it to their nest, where ant babies are lying together, hungry. Their mouths are open, waiting. One fewer cankerworm will eat the garden greens.

One spring morning, long, long ago when enormous
reptiles lived on the earth, a magnolia blossom
opened its large white petals and faced the sky.
A great beetle that had spent the night comfortably
on the flower's soft yellow center woke up, stretched
its legs, bit off a mouthful of pollen, and crawled
away. Soon a second beetle walked up the stem to
visit the blossom. Yellow pollen clung to its body
hairs, for it had just climbed out of another magnolia
blossom where *it* had been sleeping. The beetle rummaged
a bit among the stamens surrounding a cluster of pistils.

Stamens are the male parts of the flower. Each is a
small stalk crowned with an anther which holds pollen.
Pistils are the female parts of the flower. Each has
a tube leading to the ovary. The tube is called
a style and is crowned with a sticky stigma.

The stigmas curled around the beetle's body, catching
pollen as the beetle burrowed through them to the other
side of the blossom. Later the pollen that had stuck
to the stigmas would grow downward, through the style,
to join the young seeds waiting in their ovary home.
Though other insects were alive in that long-ago time,
this beetle, called Goliath, was one of the early
flower pollinators. The magnolia, one of the earliest
flowering plants, is still pollinated by beetles.

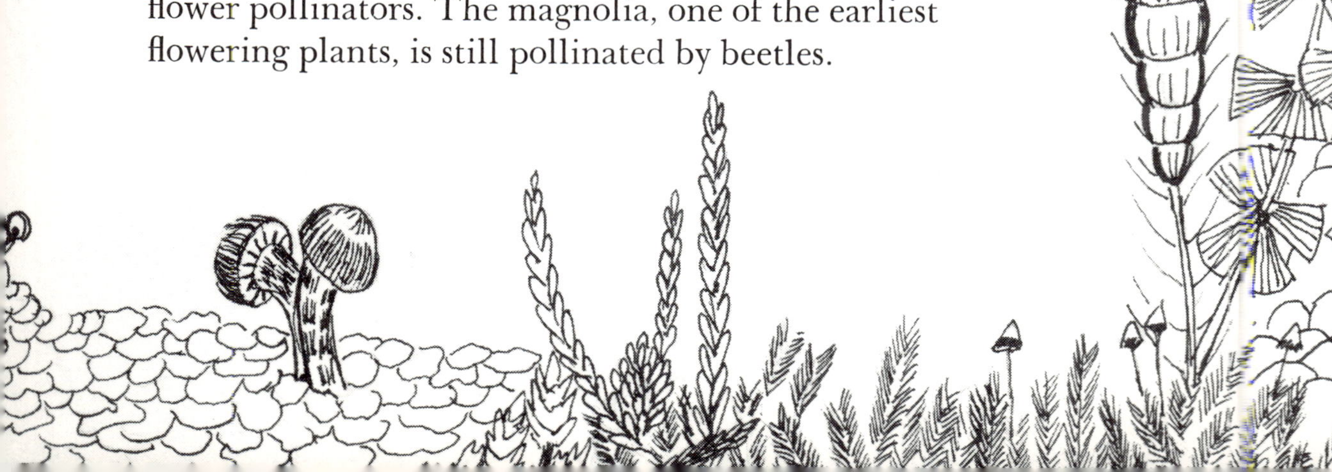

The garden is busy. Everywhere leaves are
opening. The leaf has an important job—
to breathe for the plant through its veins.
A very young caterpillar has already started
to chew holes in the willow leaf. Below the
caterpillar, on the tree trunk, a tiny wasp
called a trichogramma has found a cluster of
gypsy moth eggs which have not yet hatched.
She stops to inspect them, perches on one,
and uses her ovipositor (egg layer) to push
an egg inside it. When her egg hatches, her
infant will eat the gypsy moth egg. And so
that gypsy moth caterpillar will never hatch
to eat the willow leaves.

In another part of the garden, a young potter
wasp is looking for a hollow sumac twig left
from last year. A potter wasp is one of the
many kinds of loner wasps which live alone and
do not raise their young. The sumac twig has
exactly the right kind of space for the cells
of her nursery.

24

First an egg is attached to the inside wall of the twig.
Then the mother wasp flies away to hunt
small caterpillars and beetles, which she
catches as food for the baby wasp to eat
when it hatches. She builds a mud partition,
making her young a protected home. Again
and again she builds compartments, one for
each egg, until the last egg is safe and
sound and her job here is done.

She was once hatched from an egg on the
same kind of shelf in a sumac twig, and
when she grew up, she came outside and found
a mate. But once away from his friendly
clasp, she ever afterward stayed by herself,
a lonely loner wasp.

25

A paper wasp queen is working on her nest.
She scrapes the pulp from old soft wood and chews
and chews it into papier-mâché. She molds it into
the shape of a cup hung upside down like a chandelier.
Inside the cup she glues six-sided cells together,
fitting them like a puzzle. She lays her eggs in
these. When they hatch from the eggs, the babies
must eat. The mother hunts insects that she prepares
into tiny balls of chopped insect meat. The first young
to hatch are females. They will help their mother
enlarge the nest and will find food for the new
larvae that live in the six-sided cells. Their
mother will stay in the nest now and continue to
lay eggs. Social insects live together and help
one another.

Most social wasps feed caterpillars and other
plant eaters to their young. This gives the plants
a better chance to grow.

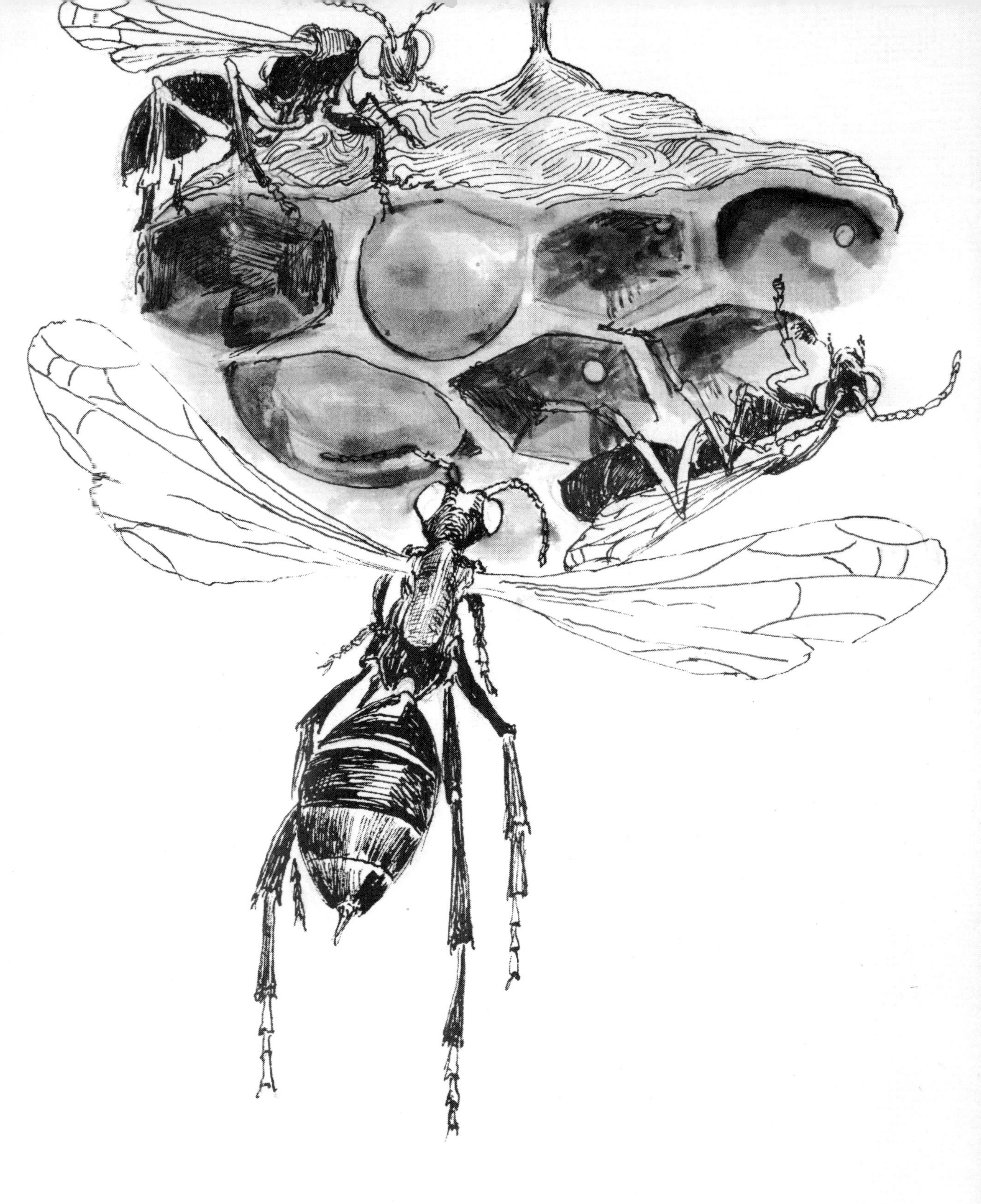

There are several kinds of bees.
Some live by themselves, building
nests in bramble bushes, making separate
little chambers where an egg is left
to hatch on a bed of pollen. Then there
are the social bees that live in families.
Their young are hatched out side by side,
with nurses taking care of them.
Food is brought for all to share.

Bees may live in colonies in the hollows
of old trees. Or they may live in hives made
by people who like bees. They make so
much honey that they are called "honeybees."

This bumblebee lives underground in a hole
she has found. Here she raises all her
children, and here they all live together.
All kinds of bees behave the same in three respects.
One is how their young are fed. Wasps
give their babies insects to eat, but bee
babies don't eat meat. The food of bee babies
is nectar and pollen, mixed into a soft bee
bread. Second, adult bees eat honey which
they make from nectar. All bees make honey.
Finally, and most important, all bees help
flowers multiply and grow.

Inside a hollow log, a young bumblebee
queen awakes from hibernation. Slowly she
creeps into the sunlight. She stretches
her wings and takes to the air. She is
looking for a nesting site. Flying low,
she spies a hole in the meadow that seems
right. It once belonged to a field mouse.

It looks like the perfect spot in which to
set up her home. Using special glands in
her body, she makes wax to build a cell for
her first eggs. Next to this she builds a
honeypot to hold the mixture of pollen,
nectar, and water that she will feed to herself
and her brood. Then she flies out to the
field to gather her provisions. And—
there's larkspur!

Certain early blooming flowers are a very
welcome sight to the first small pioneers
that come out of hiding places into the
warm sunlight. Bumblebee queens love the
nectar and pollen offered by larkspur, and
larkspur especially needs the bumblebee,
which is its only visitor. Since other bees
cannot reach the deep-lying nectar, they do
not visit. Only the bumblebee carries the
larkspur's pollen from one blossom to another.

When the young bumblebee queen returns to her nest,
she fills the honeypot with nectar and water, and pads
the cell with pollen. Then she lays several eggs
on this soft yellow bed of pollen. After sealing the
cell with wax, she sits upon it like an old mother hen.
The babies that hatch are larvae. They look like
small, fat worms. They are hungry and gobble up their
pollen bed. When it is gone, they will spin cocoons,
where they will rest as they grow and change
into bumblebees. When they wake up, their color is
gray. They drink the soup of the honeypot and rest
under their furry mother until the third day.
Then they crawl out into the bright world of spring
flowers and blossoming trees. These are the queen's
oldest daughters. They will help their mother.
Now she stays mostly in the nest to lay eggs. Her
daughters bring food for the younger bumblebees.

In late summer, new queens and males will hatch.
The queens will live through the winter sleeping
in dark protected places.

The honeybee hive has come alive!
Scouts have returned from the field, the
scent of apple blossom on their bodies.
They are dancing on the honeycomb. This
is their way of telling the hive where
they have been. Apple blossom pollen
fills the pollen baskets, which are little
brushes on their hind legs. A young bee
takes it eagerly to store in pantry cells.
One by one, workers leave the hive and
soar into the air. They will work today!
The sun shines through their wings
as they fly.

Of course, they cannot all go. Many
jobs must be done at home to prepare
the hive for the honeyflow. The youngest
bees work as nurses. They care for the
larvae that hatch from the eggs of their
queen mother. They feed and groom
the queen. They feed *royal jelly,* a
nutritious food made by nurse bees, to the
youngest grubs and nectar-and-pollen
beebread to the older ones.

36

They cover the pupa cells and
clean and prepare empty cells for new eggs.

   Certain bees are the right age to be
guards, to fight off intruders at the
entrance. When a bee finds that her body can make
wax, she helps build cells for the hive.
Soon she is ready to go outside to practice
flying. On the fifteenth day she starts to
work in the free air and sunshine. She
hunts for pollen and nectar in the flowers
of the open field and brings it home to
her family. The bees who do all this work
are daughters of the queen.

   Their brothers, the drones, do not work.
They are taken care of, for the time
will come when a new queen will hatch.
She will take a wedding flight high in the
sky. The drones will all follow and try
to catch her. One, the strongest,
will outfly his brothers to win the queen
and mate with her.

A honeybee heads for the sweet scent of
apple blossoms. She circles a part of the
orchard, flying lower to decide where to
begin. *There* is a beauty, a freshly opened
blossom, just waiting for her. The honeybee
flies to the edge of its petal and alights.
Crawling into the center, she brushes
against the stamens and is showered with
tiny grains of pollen. They stick to her
back, her legs, even her eyes. She dips her
tongue delicately—like a kiss—into the
throat of the flower. She sucks the
sweet juice, crawls out and visits another
blossom on the tree.

Here the pistil is standing tall, waiting.
Its head, or stigma, is moist and sticky.
As the bee crawls in to have her refreshment,
some of the pollen grains that cover her
are caught on the sticky stigma head of
the pistil. The pistil has a long tube
called a style. An opening in the
center of the stigma allows several of the
pollen grains to enter the style. These go
down through the style until they reach the
round, soft ovary.

The ovary holds little ovules,
seeds waiting to be fertilized by the pollen.
After the pollen reaches the ovules and
fertilizes them, the ovary will close over them
to form a fruit. This will be an apple.
The apple is a sure sign that its blossom
has been fertilized. The seed of that apple
may one day grow into another apple tree.

How warm it is today! Is it still spring?
Or has the summer come? It is that time when
wild things know the seasons best. The
meadow is so green, except for some twigs left
over from last fall. On one of these twigs
the praying mantis stopped by to lay her brood
before she died. And here the eggs spent the
winter, in tiny cells, inside a
dark brown case. They were lined up in nice, neat
corridors, protected by a stiff and sturdy wall.

Today, as morning rises, the egg case seems
alive! Little creatures struggle to get out.
Each small and squirming body, as big as a large
ant, is quite a perfect mantis, though it must
wait awhile to get its wings.

One young mantis rests a bit and stands there,
the kindest looking creature, exactly like
its mother. If it's hungry and can catch him,
it might just eat its brother.

A garden in the summer is bound to have
some pests. The mantis will go after them as
soon as it gets big enough. Right now it finds
that aphids, mealy bugs, and caterpillars,
any insects small enough, are best!

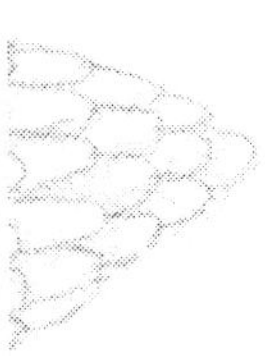

Red clover is for bumblebees.

Summer has come! The red clover is tall.
Different bees circle the top blossom.
They take turns alighting, but like window
shoppers they soon leave—all but the
bumblebee, for this is *her* kind of flower.

She lights on the reddish purple bloom and peers
into one of the many long, slender florets
of the clover head. A floret is a tiny flower.
A cluster of florets forms one blossom.
The bee's long black tongue dips into the
nectar at the bottom. Its sweet liquid is
for her alone. When she has covered one
clover head, she flies off to the next. She
visits many on the same trip and carries
pollen from one clover to another.

A damsel bug is looking for a sturdy stem.
She is ready to lay her eggs. This red clover
will do nicely. Piercing the stem with tiny
holes, one above the other, she lays an egg
in each and covers the hole with a little
oval cap. Six days later, a damsel bug
bursts the egg shell and pushes out the oval
cap with its head. After gazing out its window
for a while, it climbs through and rests
on the stem. It is hungry. An aphid is
climbing up. When it gets close, the new damsel
bug grabs it, punctures its stomach with a long
sharp beak, and drinks its blood. This way the
damsel bug protects the clover from the aphid.

43

White clover is for honeybees.

The smell of sweet white clover hangs in
the air. A honeybee sees that the white clover
field is busy with bees—and clover is best—
but one mustn't crowd, so she can't go there.
On the edge of the clover, other scents and
colors invite the honeybee. She swoops down
and lands on a laurel blossom. It's shaped
like a wheel, the stamens forming the spokes. The
anthers are held in petal pockets, like wire
springs latched in metal sockets. And when
the bee lights, the anthers fly up, throwing
their pollen onto the honeybee's body.

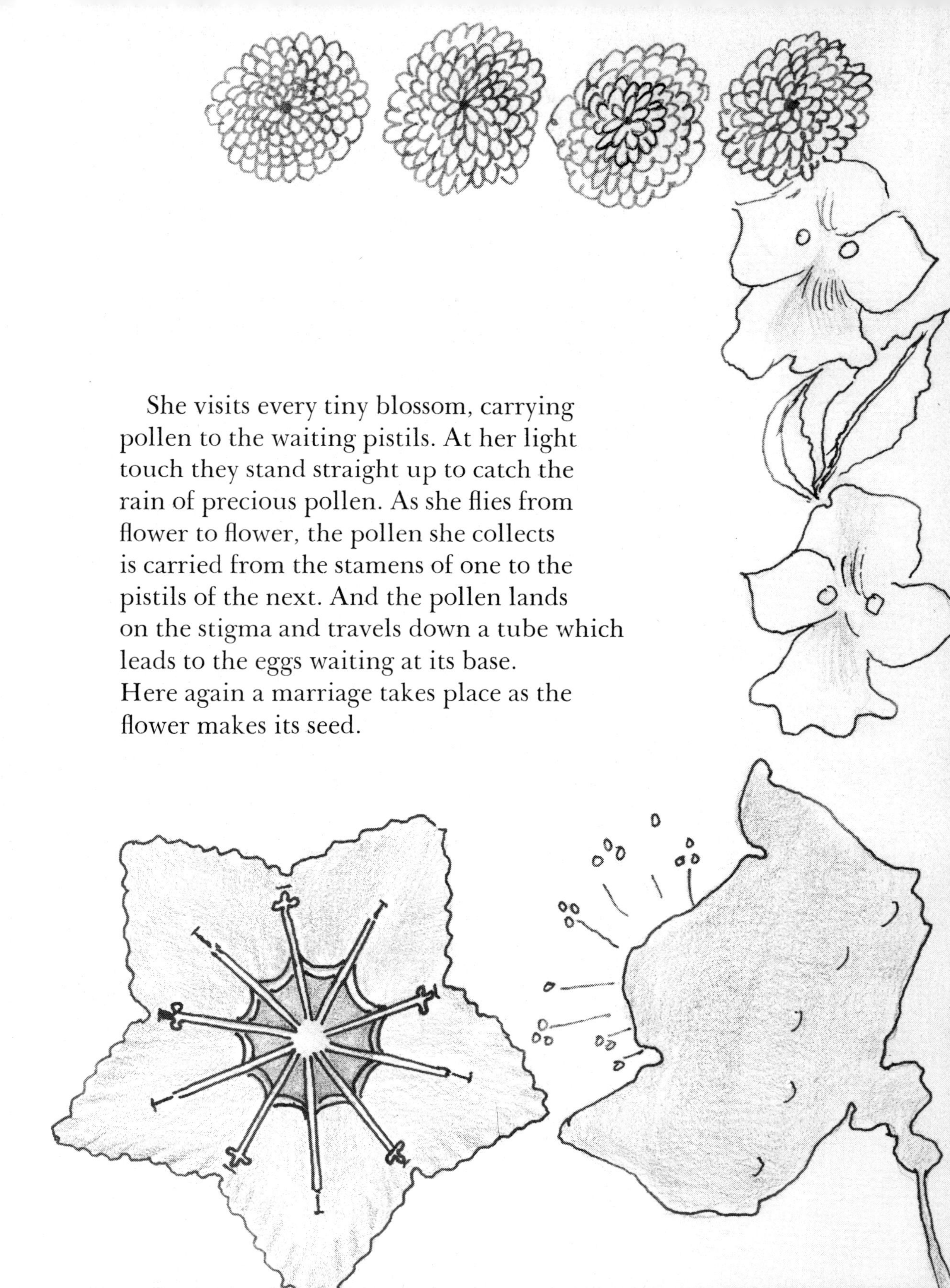

She visits every tiny blossom, carrying
pollen to the waiting pistils. At her light
touch they stand straight up to catch the
rain of precious pollen. As she flies from
flower to flower, the pollen she collects
is carried from the stamens of one to the
pistils of the next. And the pollen lands
on the stigma and travels down a tube which
leads to the eggs waiting at its base.
Here again a marriage takes place as the
flower makes its seed.

The moon has just risen on this warm night.
Cicadas sing in the trees. Blades of grass
catch the moon's light and throw shadows.
A cutworm wakes up, uncoils its body, crawls
through the dirt, and begins its breakfast of
petunia root. Frightened by the sound of
scurrying, it tries to get away by climbing the
stem of the petunia, but it is too late!
It is grabbed in the jaws of a male ground beetle!

The beetle chews the cutworm with all four
parts of his mouth. This was a good snack,
but the beetle is still hungry. He pokes
into a clump of grass near the tree, listening
for the sound of a wood-eating beetle larva.
The ground beetle will eat any insect
it can find, even another ground beetle.
At the sound of a scuttling noise, the beetle
looks at the tree trunk. There is his cousin,
a tree beetle, climbing the tree, hunting for
caterpillars. The ground beetle does not chase
his cousin. His prey must be on the ground.

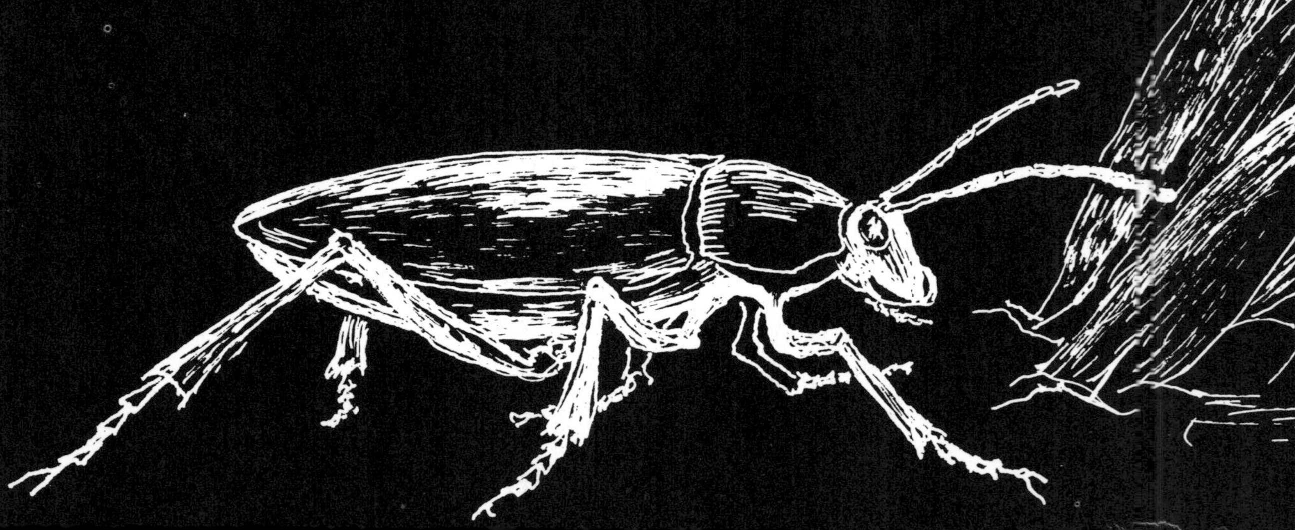

A poplar tree in the garden has become a
family tree to the aphids that live on it.
Thousands of them sit in double rows sucking leaves.
This is their food. But the tree will die if
they are not stopped. Several aphids are
having babies, so the number of insects eating
the poor poplar is growing terribly fast.

Like little jewels, the eggs of the ladybird
beetle catch the light of the sun from the
underside of a leaf. As they hatch, a tiny
dragon comes out of each one—the ladybird's
daughter or son. A fat little aphid is settling
down to suck on the juicy vein of a leaf.
Its beak is like a bent needle. One
of the ladybird larvae sees the aphid and
scrambles over to eat it.

Mother ladybird has been having a hearty meal
of aphids. When she laid her eggs on the leaf
stocked with aphids, it was her way of providing
for her young. Now they eat like their mother.
Later they will look like their mother.
The poplar tree has a chance to live!

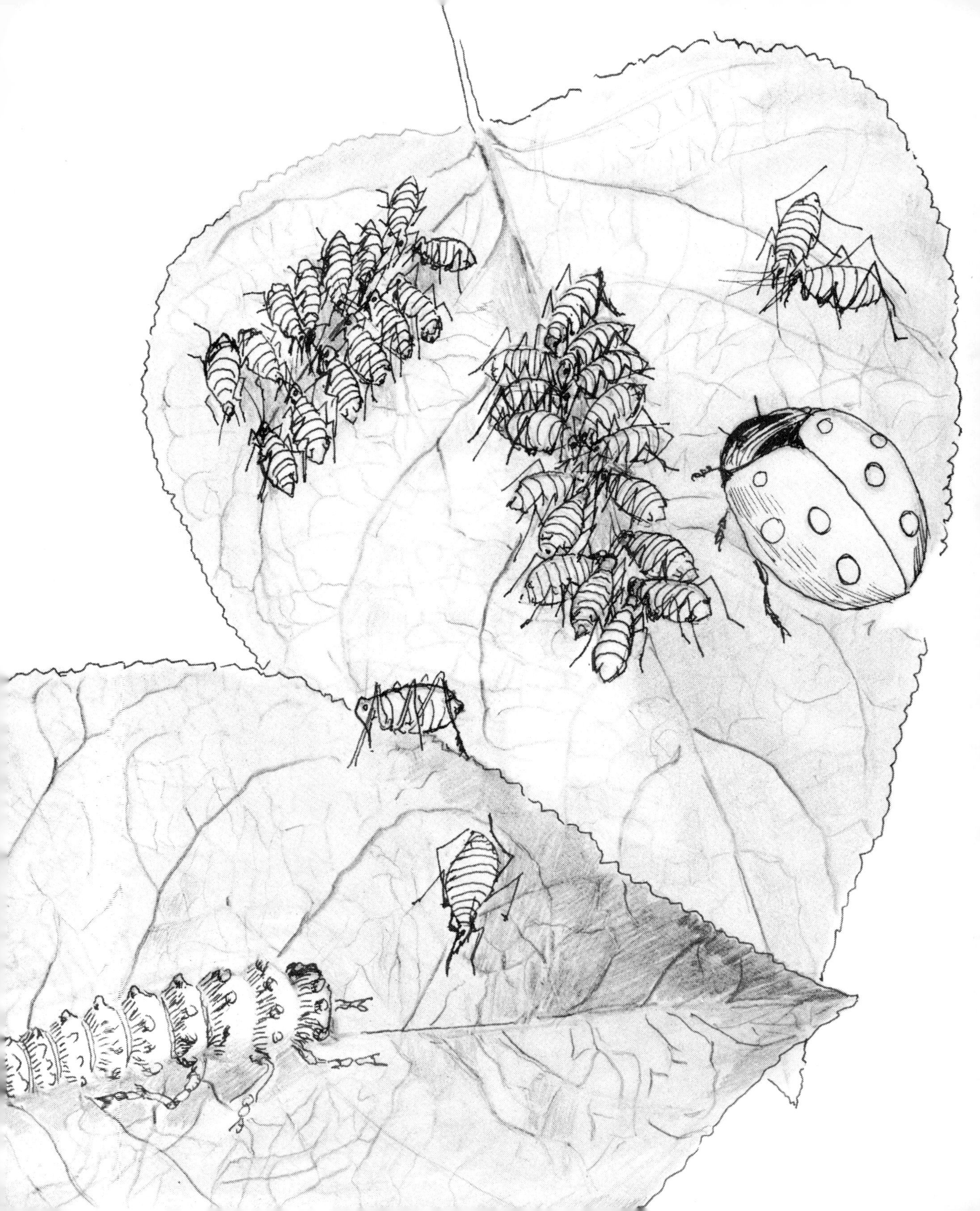

Yesterday it rained, but
today the sun is shining.
A female swallowtail butterfly
crawls out from under a stone
where she was hiding from the rain.
Today the wild bergamot blossoms will be
having many visitors.
Hummingbird moths will be there, as well as
bumblebees and flower flies, to mention
only a few.

The swallowtail flutters off to join them.
Nearly there, she sees a male swallowtail
sailing around above a section of the field. He
is guarding it for himself and some of his
female friends. She watches as another male
tries to light on a flower but is chased away.
She is not chased; only males are. Each
section of the field is guarded by a male
butterfly. They are like flying flowers, all of
different kinds and colors.

Nectar lies at the bottom of a long throat
in each of the blossom's many florets. The
butterfly has a drinking straw which she
uncoils in the face of the flower, dips into
the bottom of the floret, and sips. Her two
front legs are like delicate arms holding a
cup while she drinks from it. The male
swallowtail glides over to her and sips from
the other side of the same flower. They have
a friendly game of tag, but he must leave her
for a moment to chase another male away
from his part of the field. After doing this, he
returns to finish the game, and they fly into
the sky together.

51

The sphinx moth sweeps from nowhere out
of the dusk. The smell of tobacco blossom
fills the evening air. The moth wants to
find it and follows the scent to where
hundreds of tobacco blossoms are growing.
The moth aims for the lightest, brightest,
and biggest flower it can find, for now
it can see the blossoms as well as smell them.

52

As it approaches the blossom, the moth unfurls a
long, long tongue. Like a tube, the tongue
is carried coiled tightly under the face.
The tongue is so long that the moth does
not even have to hold on to the flower
petals. It hovers in front of the blossom like
a helicopter, its wings whirring. After
taking the nectar, it curls up its tongue,
turns quickly, crackling midair, and heads
for the deeply hidden nectar of another
flower. Some of the pollen is carried by the
tongue. And so it goes, from flower to flower.

53

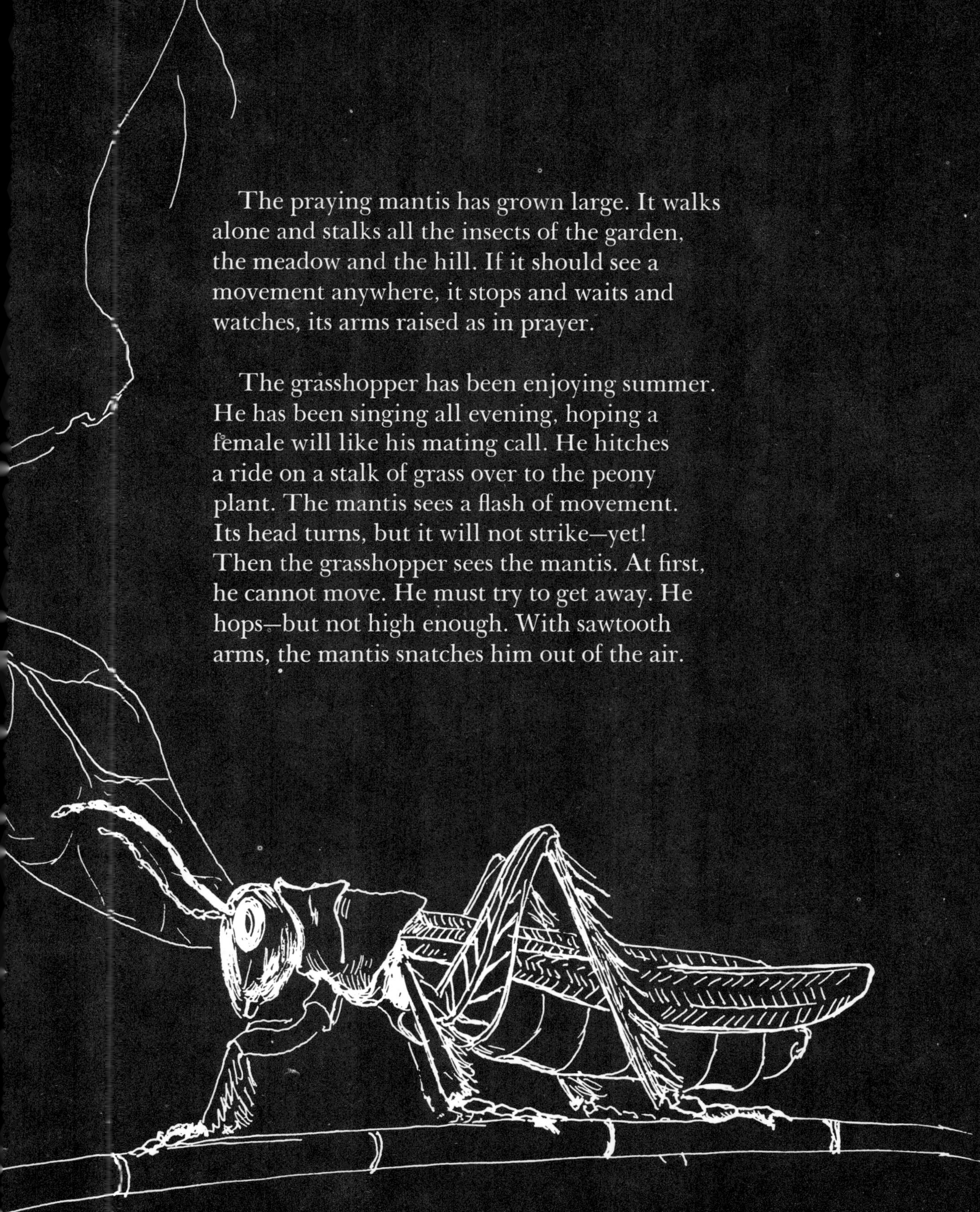

The praying mantis has grown large. It walks
alone and stalks all the insects of the garden,
the meadow and the hill. If it should see a
movement anywhere, it stops and waits and
watches, its arms raised as in prayer.

The grasshopper has been enjoying summer.
He has been singing all evening, hoping a
female will like his mating call. He hitches
a ride on a stalk of grass over to the peony
plant. The mantis sees a flash of movement.
Its head turns, but it will not strike—yet!
Then the grasshopper sees the mantis. At first,
he cannot move. He must try to get away. He
hops—but not high enough. With sawtooth
arms, the mantis snatches him out of the air.

The honeybee is very excited. Now it's fall and
the leaves are turning. She is making her first flight
to the fields. She will be one of her queen
mother's attendants through the winter months.
She must do her part to get the hive ready for
the frightening freezing time ahead. Many bees
will die of the cold. The honey is their fuel
for the winter. When the bees eat, their
bodies become warm, and they cluster in a ball,
with their queen at the center to keep
her warm.

A tall goldenrod waves and beckons in the field.
The female honeybee lands on its yellow ramp.
There are so *many* refreshing flowers! Each tiny
floret has its own nectary, which makes the nectar.
She sips from them gratefully. Phlox and asters
grow nearby, but she visits only one kind of
flower at a time. When she has had her fill of
goldenrod, she will fly home. Her body will
start making honey from this nectar while she
is flying. She has done good work for the hive
today.

Some of the ant leaders are getting the
underground chambers ready for cold weather.
Winds from the north have been blowing the
red, yellow, brown and purple leaves into piles.
The ant colony must have enough food stored
below to keep from starving until spring warms
the earth again. Some important guests will spend
the winter with the ants in their underground
home. These are aphids who have mated and will
soon be laying eggs. The aphid eggs will be
tended by ants until they hatch in the spring.

The ants must dig new storage bins
underground for their winter food. Digging
tunnels, they help the earth by loosening it and
mixing it with air. This will make it easier for
plants to grow next year.

Today, a long line of ants is following
its leaders over the hill. Each ant
carries something in its mouth: a dead insect,
a bit of decayed wood, aphids, aphid eggs.
They march in single file to the entrance of
their home and file down the steep passage.
The last few ants in line close the opening
with dirt.

Now the queen will stop laying eggs.
Those ant larvae already hatched will sleep
more than they will eat. And they will grow
very, very slowly. All the ants'
movements will slow down. The whole colony
will rest quietly until spring.

After mating, the praying mantis eats even
more than she did before. She even eats her
mate in order to nourish the fertilized eggs.
Then she must find a place to lay her egg mass.

A few feet above the ground she selects a
twig. Hanging face downward, she holds
on with her middle and hind legs. Her abdomen,
the hind section of her body, is now swollen
and looks like a full plastic tube.
Its pointed tip opens and divides into two ends
which have the shape of mixing spoons. From
the opening, a ribbon of heavy gray liquid is
pushed out, then beaten into foam by the two
mixing spoons. The work takes about two hours.
And all the while, eggs are coming out and taking
their place in the frothy mass. The abdomen
of the mantis seems to work by itself. Mother
mantis never looks around to check up on the job
it is doing, but her egg mass will have everything
that her eggs will need to be safe all
winter. It has tiny passageways for the young
mantises to come through when they hatch in
the spring, and there is one hallway in the
very center through which they all travel
to find the light of day.

The mass becomes a case with heavy walls
to protect the eggs from snow and rain.
But the main protection which the mantis
gives her eggs is the air beaten into the
egg case. This air, trapped inside tiny
layers of silk, stops the cold from entering,
just as the attic on a house does.

When all the eggs are laid, the mantis
walks away. Soon she will die, but from this
hardened egg mass new mantises will next
year carry on the work of the garden
guardian.

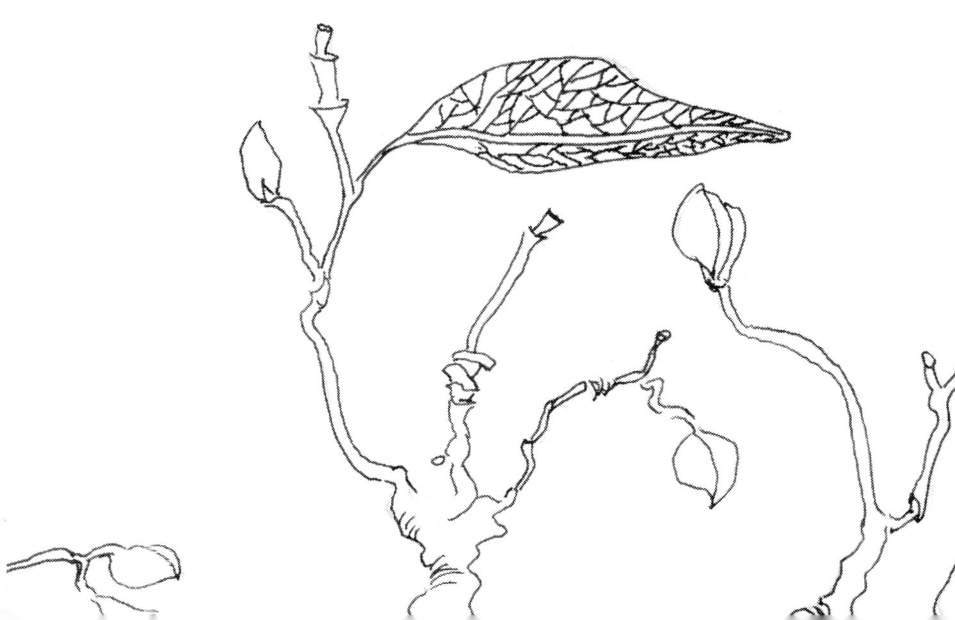

The larkspur seed which was pollinated last spring by the bumblebee has finally fallen to the ground. At first it stays in a case with many other seeds. Then, the rain opens the case and draws the seed into the ground. It will lie there all winter.

Three weeks ago a male and a female grasshopper mated. Autumn winds now warn the world of frost. The female grasshopper, whose mate still sings in the grasses, hops across the meadow looking for a place to lay her eggs. That spot of bare ground next to the withered larkspur will do nicely. With the tip of her ovipositor, the mother drills a hole in the ground.

It is hard work, but she has a good drilling tool. The two pointed ends of her abdomen open and close to loosen the dirt as she pushes deeper and deeper. When her whole abdomen is buried, she begins to fill the hole with eggs. She does a neat job, making exactly four rows with seven eggs in each.

She arranges them carefully so that if any of the lower eggs hatch first, those young grasshoppers will be able to escape from underground. When the mother has emptied herself of eggs, she fills the top of the hole with a sticky fluid that will dry into a shell around the eggs. Now it is an eggpod. Before long the mother and father grasshoppers lose interest in eating, become too weak to hop, and die.

Inside each egg, magic has happened. The tiny cells, one from the mother, one from the father, have joined. Together, they will grow into a new grasshopper. But for now, the egg will sleep next to the larkspur seed.